THE FLOWER OF LIFE PATTERN

THE SACRED VESICA PISCIS REVEALS THE ANATOMY OF THE FLOWER OF LIFE PATTERN AND A DISCOVERY WITHIN IT

THE FLOWER OF LIFE PATTERN

THE SACRED VESICA PISCIS REVEALS THE ANATOMY OF THE FLOWER OF LIFE PATTERN AND A DISCOVERY WITHIN IT

BRENT CROWDER

Copyright © 2025 by Brent Crowder

All rights reserved. No part of this publication may be reproduced, distributed, or transmitted in any form or by any means, including photocopying, recording, or other electronic or mechanical methods, without the prior written permission of the copyright owner and the publisher, except in the case of brief quotations embodied in critical reviews and certain other noncommercial uses permitted by copyright law. For permission requests, write to the publisher, addressed "Attention: Permissions Coordinator," at the address below.

CITIOFBOOKS, INC.
3736 Eubank NE Suite A1
Albuquerque, NM 87111-3579
www.citiofbooks.com

Hotline:	1 (877) 389-2759
Fax:	1 (505) 930-7244

Ordering Information:
Quantity sales. Special discounts are available on quantity purchases by corporations, associations, and others. For details, contact the publisher at the address above.

Printed in the United States of America.

ISBN-13:	Softcover	979-8-90124-308-4
	eBook	979-8-90124-309-1

Table of Contents

1 Introduction 1
2 The Sacred Vesica Pisces Circles and the Seed of Life 2
3 Flower of Life Concentric Increments 9
4 The Flower of Life Image and Pattern 16
5 The Flower of Life Pattern, Discovery of Whole Number Square Root Radii 19
6 Expanding the Flower of Life Pattern by Symmetry 22
7 Tabulation of Hexagons, Increments, and r Squared Values 24
8 Observations Regarding Flower of Life Pattern Graphical and Tabular Characteristics 26
9 Graphical Derivation of "Missing" Whole Number r Squared Values 28
10 Expansion by Symmetry 30
11 Summary 32
References 33
ABOUT THE AUTHOR 34

Chapter 1

Introduction

In "Sacred Geometry", the vesica piscis evolves via a point, centering the resulting first circle, and then an equal radius second circle compassed from any point on the first circle. The reason for its sacredness has been documented, but not often enough divulged, recognized, or credited to those who first acknowledged what arguably makes the vesica piscis sacred. Then, the six additional circles, equally incremented around the first one, creating the pattern known as the "Seed of Life" evolves, which, when extended via incrementally outward circles centered about the very first circle, becomes the flower of life pattern. This pattern and extensions thereof will be described and analyzed to discover and reveal the whole number square roots inherently graphed within, and they will be tabularized in a predictable, expandable, and surprisingly simple manner.

Chapter 2

The Sacred Vesica Piscis Circles and the Seed of Life

Unless stated otherwise, assume all circles to have a radius of 1, as determined by the initial compass setting.

Draw a circle with a compass (a real one, or via computer aided drafting techniques). The second circle, constructed without changing the radius is chosen to pivot from anywhere on the first circle to form what is known as the vesica piscis. In Figure 1, below, the second (upper) circle was pivoted from the very top of the first circle.

Figure 1

First and Second Circles to Form a Vesica Pisces

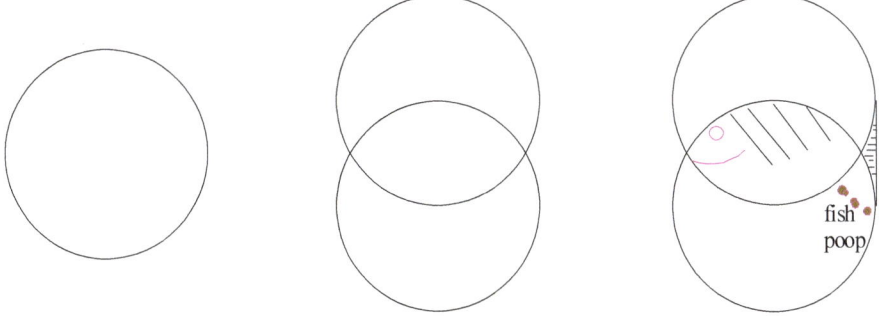

Some geometric properties of the vesica piscis become evident by adding five more circles around one circle until six "flower petals" emerge in what is commonly known as the seed of life as shown in Figure 2.

Figure 2

Six Circles Around One to Obtain the Seed of Life

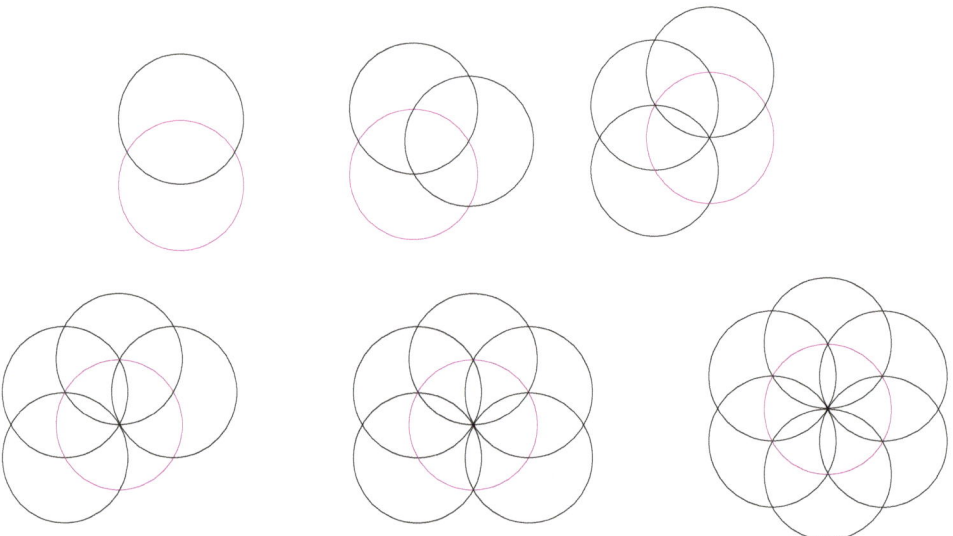

The Seed of Life

The seed of life, including a detail of a vesica piscis, within, is shown in Figure 3. The vesica piscis is a symbol made from two circles of the same radius, intersecting in such a way that the center of each circle lies on the circumference of the other. According to Wikipedia[1], the name literally means the bladder of the fish in Latin. In the Christian tradition, it is a reference to Christ, as in ichthys[2]. It appears in the first proposition of Euclid's Elements, where it forms the first step in constructing an equilateral triangle using a compass and straight edge. Wikipedia notes the ratio 265:153 as a good approximation of the square root of 3 as determined by Archimedes, who like another Greek mathematician, Pythagoras, lived a few hundred years before Jesus Christ. The Bible's book of John describes Jesus, after being raised from the dead, instructing, from the beach, his empty netted fishing disciples to cast the nets on the right side of the boat, and they drew in, per

John 21:11, 153 fish. "John lived and wrote his Greek Gospel in the Greek city of Ephesus. He knew they falsely believed the Gospel of Jesus was foolishness and in contradiction to their wisdom." "The Apostle, John was clearly appealing to those who would see the relevance of this mention of 153 in the ratio 265/153 as the best approximation of the measure, root 3, of the fish." [3]John 21:11: "So Simon Peter went aboard and hauled the net ashore, full of large fish, 153 of them; and although there were so many, the net was not torn." [4]"The detail of 153 fish seems too specific to be insignificant. Did John have a special meaning by the reference of a 153 fish?" The following are also excerpts from Ref. 3:"Above all else the Greeks esteemed Mathematics, Wisdom and Philosophy. Their greatest mathematician was Archimedes. In his Measurement of a Circle he calculates the value of π, Pi. It was his most influential work. His calculations included the ratio of 265:153 which represents the most accurate value of $\sqrt{3}$ that can be expressed by using small whole numbers. The unusual number of 153 figured most prominently in that work. The Greeks would have recognized that John was alluding to all wisdom, but to especially that of the Greeks as represented by Archimedes by using that number 153." 265 is to 153 as root 3 is to 1, the measure of the fish.

The seed of life provides a clear perspective for determining the inner dimensions of the vesica piscis because of the symmetry of the circle radius length flower petals. The three segments of triangles connecting every other petal are obviously of equal length forming a triangle inscribed into a unit radius circle. The horizontalbisector creates 30 – 60 – 90-degree triangles, allowing the length, C (= 2a), of the equilateral legs to be calculated.

Figure 3

Vesica Piscis Internal Dimensions

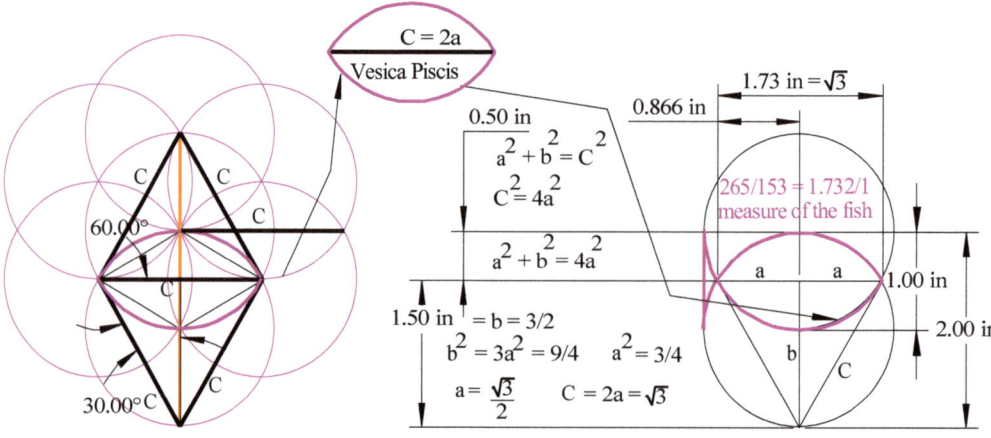

Note the above horizontal "C" equals the other Cs because the Cs connect every other "petal" of the center circle as well as connecting the intersection points of a vesica piscis. The petal-length rhombus, an equilateral parallelogram consisting of four 30-60-90 triangles, inside the vesica piscis with edge lengths equal to the unit circle radius, and the length of the vesica piscis is the square root of 3.

The two-dimensional seed of life introduced in Figure 2 can also be visualized in three dimensions as shown in Figure 4, in the form of a "cell cluster" known as the "Egg of Life", which is representative of biological cell division at an early stage of development.

Figure 4

Two Dimensional Seed of Life to Three Dimensional Egg of Life

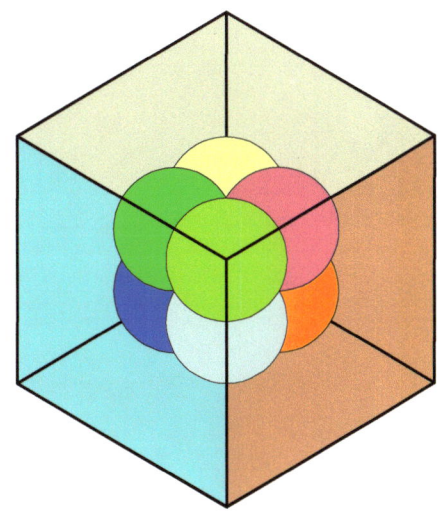

The Seed of Life

Egg of Life

Before proceeding to the detailed geometry characteristics of the flower of life pattern, the self evident ARCHitechtural and religious significance of the vesica piscis is depicted by Figure 5.

Figure 5
The Vesica Pisces as a Sacred Arch

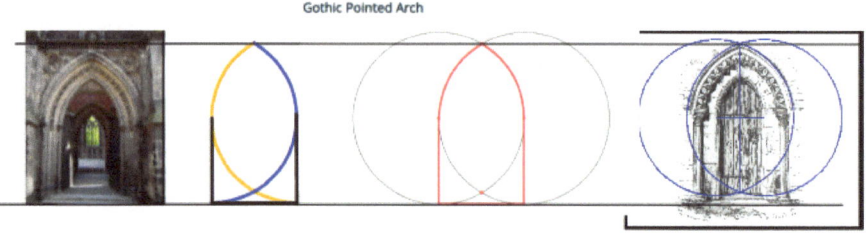

See references 5, 6, and 7 as sources for the items in the figure above..

Chapter 3

Flower of Life Concentric Increments

The flower of life results from expanding radially outward with additional circles centered on circle intersections. Drawing the additional circles requires careful placement of the compass at each point of intersection of circles. No measurement nor use of a straight edge is required. While drawing circles with a compass, the hexagonal symmetry of pattern about the center of the pattern emerges, so that one might naturally tend to draw circles incrementally outward along hexagonal segments of hexagons of increasing concentric distances from the center of the pattern.

The following discussion focuses on discovering, (or re-discovering) the radial distances of circles that occur when the circles are identified by common distances from the center to determine the radius of each concentric circular ring about the center of the pattern.

Within the seed of life, the length of each of the six flower petals is easily recognized as 1 from the method of its creation as shown in Figures 2 and 6. The next increment, increment 1, consists of the six red increment 1.0 radius circles centered at each outer vesica piscis intersection created by the seed of life circles. These circles are centered at radius square root of 3 from the center of the seed of life as shown by the black horizontal of length C in Figure 3 and in increment 1 of Figure 6, below.

Figure 6

Flower of Life
Increments 1 - 3

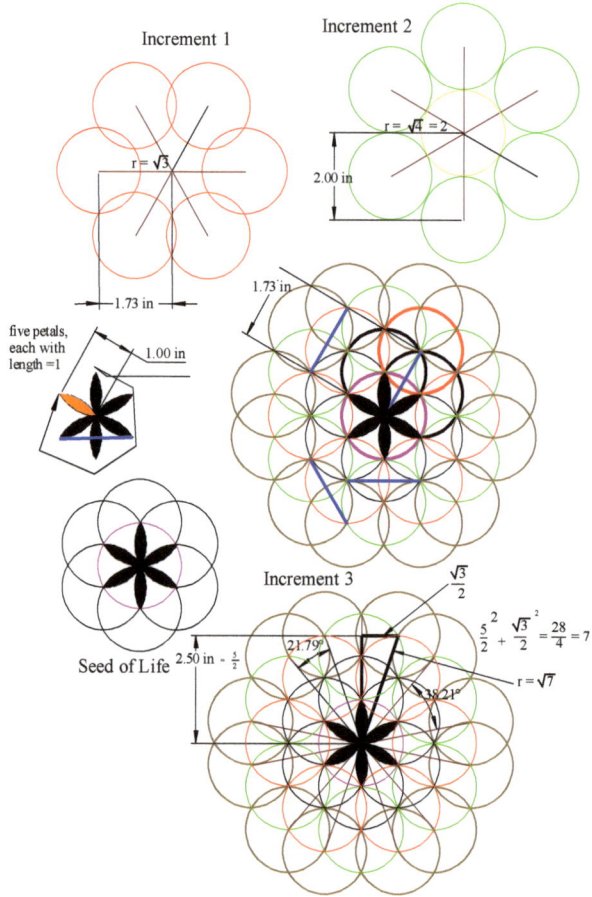

The flower petals perpendicular to the blue lines in increment 3 also serve as a sort of measuring grid for determining the radii from center of the flower of life pattern to the sets of circles arranged concentrically about the center of the pattern.

Increment 1

The set of red circles at the next larger radius than the six outer black seed of life circles is labeled "Increment 1". The radius from the pattern center to the centers of increment 1 circles is the square root of 3, which is the length of the dark blue lines in the right middle graphic of Figure 6. The blue line, dimensioned with the length 1.73 inches, should be recognized as having the length, C, from Figure 3. The dark blue line length is the long axis of the Vesica Pisces formed by the two heavy black circles and constitutes the radius from the center of the pattern to the center of the red increment 1 circles.

Increment 2

The set of six green circles at the next larger radius than the six red increment 1 circles is labeled "Increment 2". These circles are tangent to each other as well as being tangent to the center circle of the whole pattern. The radius from the pattern center to the centers of the green, increment 2 circles is the square root of 4.

Increment 3

The set of 12 light blue circles at the next larger radius than the six green increment 2 circles is labeled "Increment 3". The radius of increment 3 circles is shown as the heavy black hypotenuse defined by its heavy black vertical and horizontal legs. The vertical leg length is 2.5. The horizontal leg length is ½ the square root of three, so the hypotenuse, the square root of 7 is the radius to the center of the increment 3 circles.

Increment 4

The set of 6 black circles at the next larger radius, than the 12 light blue increment 3 circles, is labeled "Increment 4" in Figure 7, below. The radius (from the pattern center) of increment 4 circles is easily determined to be 3, which is the square root of 9.

Increment 5

The set of 6 red circles at the next larger radius than the six black increment 4 circles is labeled "Increment 5". The radius of increment 5 circles is shown as the heavy black hypotenuse defined by its heavy black vertical and horizontal legs. The vertical leg length is 3. The horizontal leg length is the square root of three, so the radius of increment 5 circles is the square root of 12. So, to review, the increments, so far have been the square roots of 1, 3, 4, 7, 9, and 12. What will be the next larger radius?

Increment 6

The set of 12 blue circles at the next larger radius than the six red increment 5 circles is labeled "Increment 6". The radius of increment 6 circles is shown as the heavy black hypotenuse defined by its vertical and horizontal legs. The vertical leg length is 7/2. The horizontal leg length is ½ the square root of three, so the radius of increment 6 circles is the square root of 13.

Increment 7

The radius of increment 7 circles is the square root of 16, or 4. This set of 6 outer green circles is shown again in Figure 7 along with the green increment 2 circles directly between the center circle and the increment 7 circles. These thirteen circles, including the center circle, is known as "The Fruit of Life".

Figure 7

Flower of Life
Increments 4 - 7

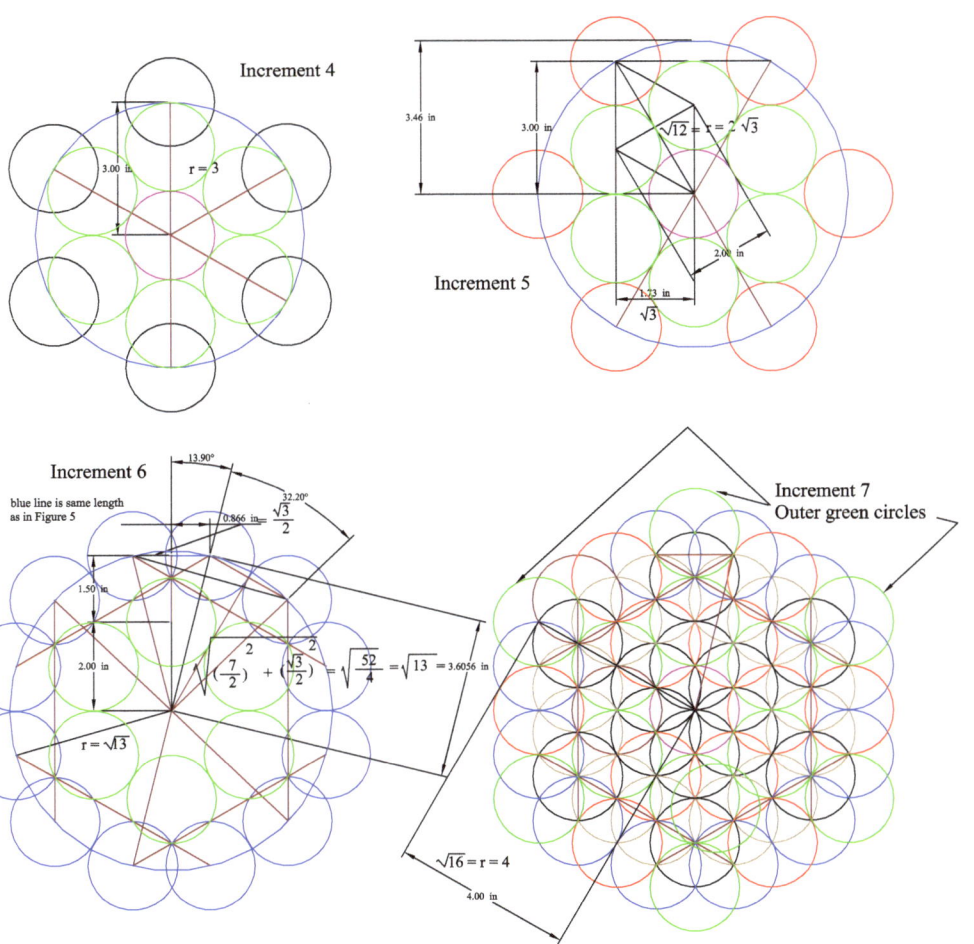

Figure 8

Flower of Life
Increments 7 - 9

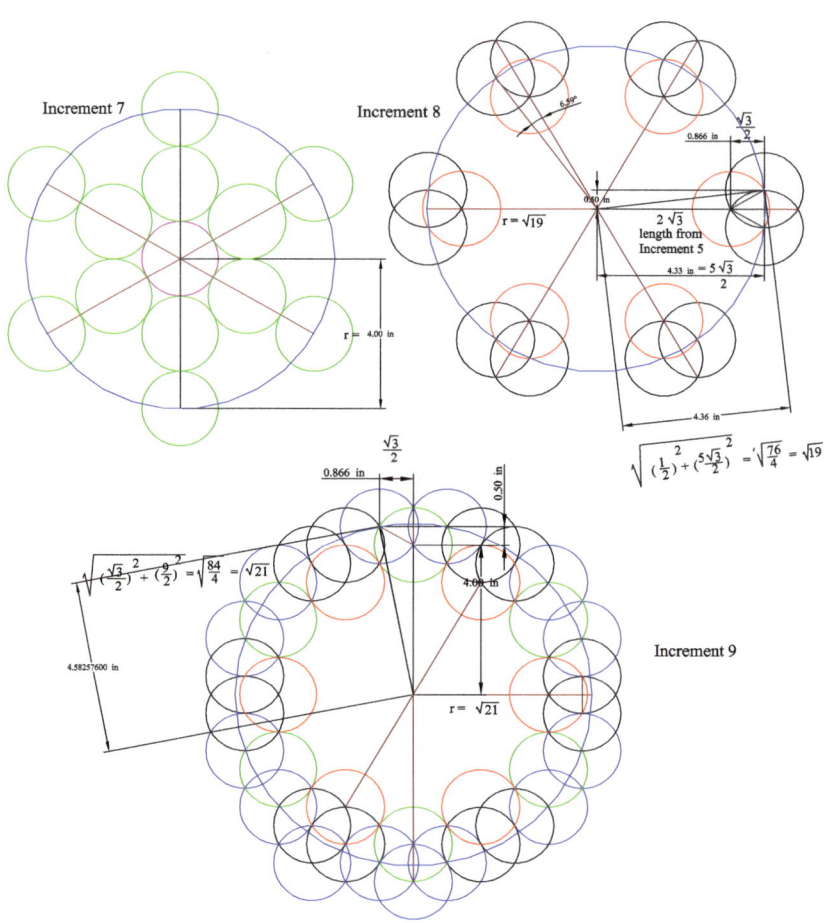

Increment 8

The set of 12 outer black circles at the next larger radius than the six green increment 7 circles is shown to have radius of the square root of 19.

Increment 9

The set of 12 outer blue circles at the next larger radius than the twelve black increment 8 circles is shown to have radius of the square root of 21.

Chapter 4

The Flower of Life Image and Pattern

The right image of Figure 9 is more extensive than the conventional image depicted as the double circle containing the conventionally recognized flower of life pattern on the left side.

Figure 9

 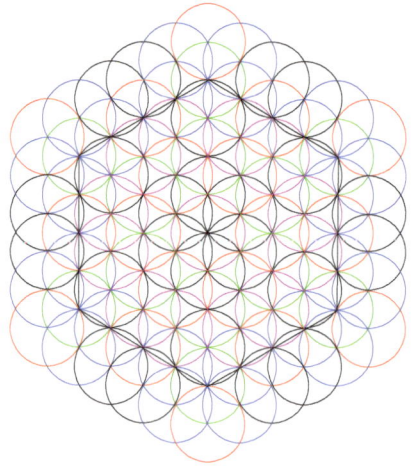

The double circled image and pattern within the double circle is called "The Flower of Life".

The circles of the flower of life can be drawn along the segments of hexagons The inner hexagon has the six circles of the seed of life, the next outer 6 segments, successively consisting of 2, 3, 4, and 5 circles (not counting the corners twice), so that each sextant consists of 1+2+3+4+5 = 15 circles, thus resulting in the above (right side) flower of life consisting of 91 circles, including the center circle (6 x 15 + 1 = 91). The double circled pattern on the left in Figure 9 has 19 complete circles. One must draw an additional 36 concentrically and incrementally evolving unit circles as determined by circle intersections to create the arcs that complete double circled image. Then, arcs outside the three-frequency pentagon are trimmed away to leave the lower right image of Figure 10.

Figure 10

Progression From 19 Flower of Life Circles to Flower of Life Circle

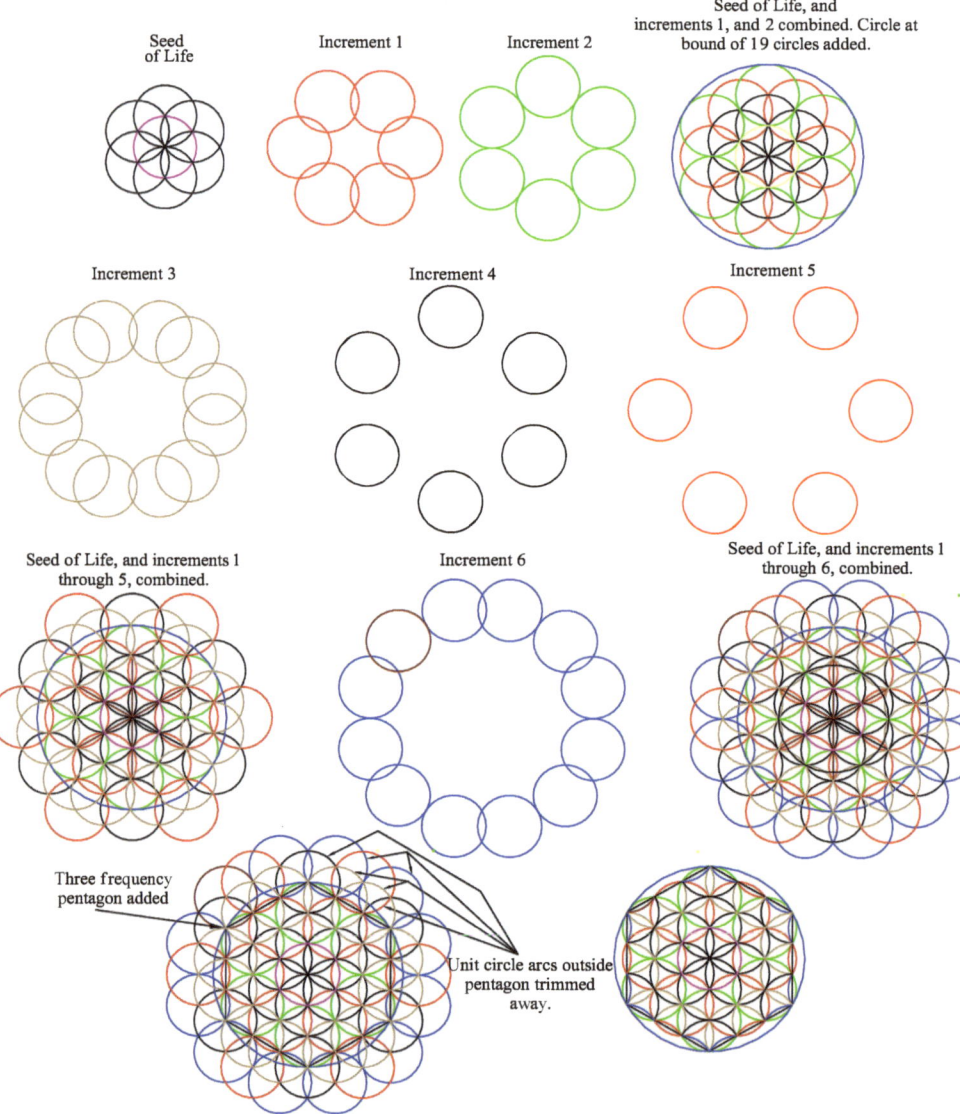

Chapter 5

The Flower of Life Pattern, Discovery of Whole Number Square Root Radii

Upon observationof the radii from the center of the flower of life pattern to the centers of its circles, it becomes intuitively apparent that each and every radius from the pattern center to the center of any of the pattern's circles is the square root of a whole number. Figure 11, below shows that extension of the pattern occurs as equally spaced concentric hexagons, incrementally spaced, with edges ½ the square root of 3 apart. The numbers in the shaded triangles are the squares of the distances from the center. By symmetry, the lower 30 - degree arc of the bottom right hand 60 - degree arc is the means of showing further expansion of the pattern as shown in Figure 12. The flower of life pattern, as shown, being created with concentric hexagons creates a grid with ½ root 3 spacing on the x axis and spacing on the y axis of 1. The distance squared from the center to any pattern unit circle is easily obtained by the Pythagorean theorem. For example, in the seventh hexagon: $36.75 + .5^2 = 37$. The figures below are intended to show r squared values rather than (x,y) values so that whole numbers corresponding to the sum of the squares are presented. A second example is shown for (-6.06218, 3.5), resulting in the r square value corresponding to: $-6.06218^2 + 3.5^2 = 36.75 + 12.75 = 49$.

Also, of primary significance, as shown in Figure 6, the petal length of 1 being used as a natural "Measuring Stick" can be used to most conveniently count seven petals for a distance of 7 (and 7 x 7 = 49) from the pattern center to the lower right vertex of the seventh hexagon.

This technique is adopted to allow the whole number squared values associated with concentric pattern unit circles to easily be identified, as well as revealing and using a tabular method, not requiring so many circles to be drawn to appreciate the "Whole Number" - "Square Root" relation of the pattern including the ability to graphically draw the square root lengths associated with the whole numbers that are not explicitly inherent in connecting the centers of the flower of life pattern circles.

Figure 11

Expansion from the Sixth to the Ninth Concentric Hexagon

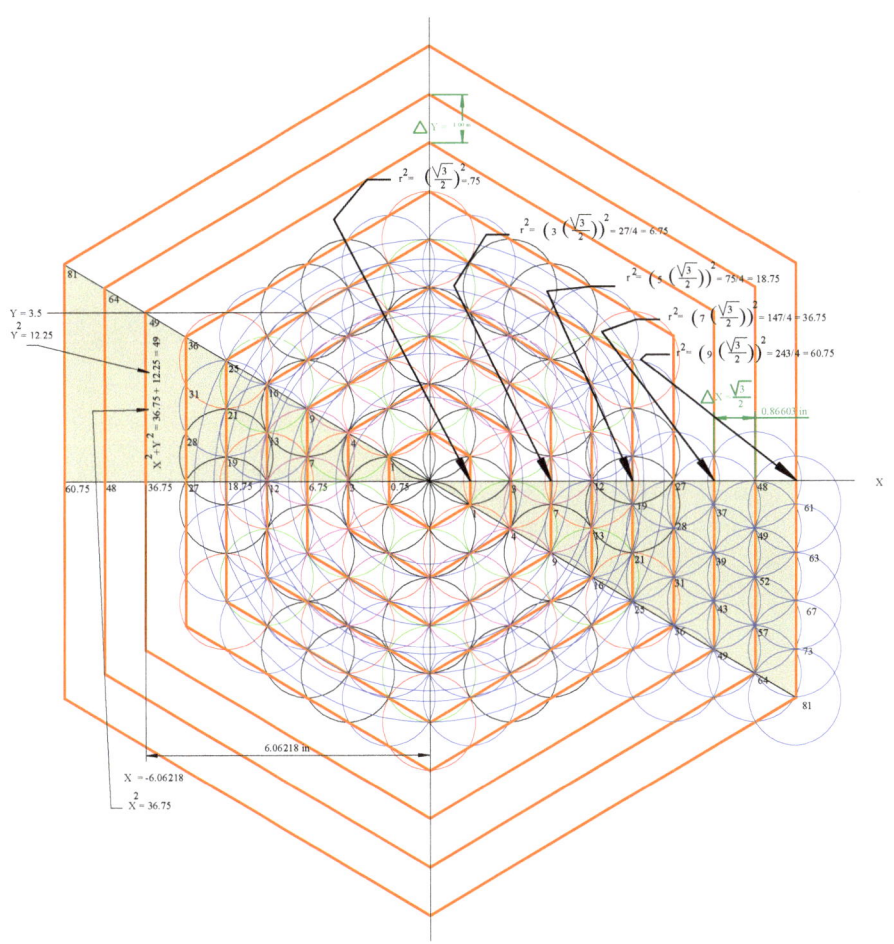

Chapter 6

Expanding the Flower of Life Pattern by Symmetry

Figure 12, below expands the pattern to the thirteenth hexagon based on expansion of the lower right shaded triangle in Figure 11, above.

Figure 12
Expansion to the Thirteenth Hexagon

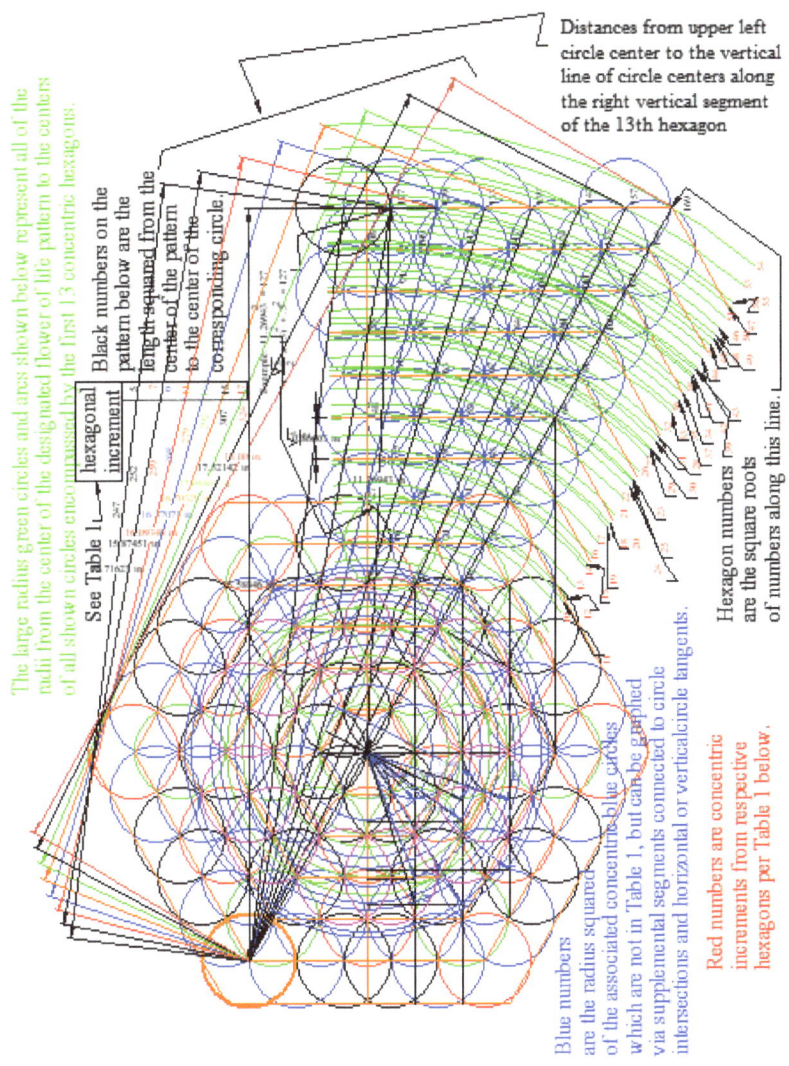

Chapter 7

Tabulation of Hexagons, Increments, and R Squared Values

Table 1, below, shows the radii (squared) of the centers of the green circles from the pattern center. For example, consistent with Figure 12, the 18 easily countable petal lengths from the longest lower left to upper right dimension results in the 18 x 18 r-square value of 324. The distance from the upper left circle center to the next circle, incrementally just above the bottom right hexagon vertex is shown as 17.52142, which, when squared is 307 and the hexagonal increment along the 18th hexagon centered out from said upper left circle is 324 − 307 = 17. Also note that "Pythagorean" squares of the x(15.59) and y(-8) coordinates correspond to the r squared value of 307.

Table 1

Description	Hexagon No.	center Increment	Intercepts X	Intercepts Y	radius squared	hex incrmnt	No. of circles
Flower of Life Concentric Characteristics							
center	0			0	0		1
seed of life	1		0.866	-0.5	1		6
	2	1	1.732	0	3		6
	2	2	1.732	-1	3	0	6
	3	3	2.598	-0.5	7		12
	3	4	2.598	-1.5	9	2	6
fruit of life	4	5	3.464	0	12		6
	4	6	3.464	-1	13	1	12
	4	7	3.464	-2	16	3	6
	5	8	4.33	-0.5	19		12
	5	9	4.33	-1.5	21	2	12
	5	10	4.33	-2.5	25	4	6
	6	11	5.196	0	27		6
	6	12	5.196	-1	28	1	12
	6	13	5.196	-2	31	3	12
	6	14	5.196	-3	36	5	6
	7	15	6.062	-0.5	37		12
	7	16	6.062	-1.5	39	2	12
	7	17	6.062	-2.5	43	4	12
	7	18	6.062	-3.5	49	6	6
	8	19	6.928	0	48		6
	8	20	6.928	-1	49	1	12
	8	21	6.928	-2	52	3	12
	8	22	6.928	-3	57	5	12
	8	23	6.928	-4	64	7	6
	9	24	7.794	-0.5	61		12
	9	25	7.794	-1.5	63	2	12
	9	26	7.794	-2.5	67	4	12
	9	27	7.794	-3.5	73	6	12
	9	28	7.794	-4.5	81	8	6
	10	29	8.66	0	75		6
	10	30	8.66	-1	76	1	12
	10	31	8.66	-2	79	3	12
	10	32	8.66	-3	84	5	12
	10	33	8.66	-4	91	7	12
	10	34	8.66	-5	100	9	6
	11	35	9.526	-0.5	91		12
	11	36	9.526	-1.5	93	2	12
	11	37	9.526	-2.5	97	4	12
	11	38	9.526	-3.5	103	6	12
	11	39	9.526	-4.5	111	8	12
	11	40	9.526	-5.5	121	10	6
	12	41	10.39	0	108		6
	12	42	10.39	-1	109	1	12
	12	43	10.39	-2	112	3	12
	12	44	10.39	-3	117	5	12
	12	45	10.39	-4	124	7	12
	12	46	10.39	-5	133	9	12
	12	47	10.39	-6	144	11	6
	13	48	11.26	-0.5	127		12
	13	49	11.26	-1.5	129	2	12
	13	50	11.26	-2.5	133	4	12
	13	51	11.26	-3.5	139	6	12
	13	52	11.26	-4.5	147	8	12
	13	53	11.26	-5.5	157	10	12
	13	54	11.26	-6.5	169	12	6
	14	55	12.12	0	147		6
	14	56	12.12	-1	148	1	12
	14	57	12.12	-2	151	3	12
	14	58	12.12	-3	156	5	12
	14	59	12.12	-4	163	7	12
	14	60	12.12	-5	172	9	12
	14	61	12.12	-6	183	11	12
	14	62	12.12	-7	196	13	6
	15	63	12.99	-0.5	169		12
	15	64	12.99	-1.5	171	2	12
	15	65	12.99	-2.5	175	4	12
	15	66	12.99	-3.5	181	6	12
	15	67	12.99	-4.5	189	8	12
	15	68	12.99	-5.5	199	10	12
	15	69	12.99	-6.5	211	12	12
	15	70	12.99	-7.5	225	14	6
	16	71	13.86	0	192		6
	16	72	13.86	-1	193	1	12
	16	73	13.86	-2	196	3	12
	16	74	13.86	-3	201	5	12
	16	75	13.86	-4	208	7	12
	16	76	13.86	-5	217	9	12
	16	77	13.86	-6	228	11	12
	16	78	13.86	-7	241	13	12
	16	79	13.86	-8	256	15	6
	17	80	14.72	-0.5	217		12
	17	81	14.72	-1.5	219	2	12
	17	82	14.72	-2.5	223	4	12
	17	83	14.72	-3.5	229	6	12
	17	84	14.72	-4.5	237	8	12
	17	85	14.72	-5.5	247	10	12
	17	86	14.72	-6.5	259	12	12
	17	87	14.72	-7.5	273	14	12
	17	88	14.72	-8.5	289	16	6
	18	89	15.59	0	243		12
	18	90	15.59	-1	244	1	12
	18	91	15.59	-2	247	3	12
	18	92	15.59	-3	252	5	12
	18	93	15.59	-4	259	7	12
	18	94	15.59	-5	268	9	12
	18	95	15.59	-6	279	11	12
	18	96	15.59	-7	292	13	12
	18	97	15.59	-8	307	15	12
	18	98	15.59	-9	324	17	6
Total							1033

Chapter 8

Observations Regarding Flower of Life Pattern Graphical and Tabular Characteristics

Figures 6, 7, and 8, above addressed center increments of 1 through 9, so, in order to avoid additional clutter to an already "busy" Figure 12, the last increment, 10, on the vertices of hexagon 5, is the first center increment shown (in red), followed by increment 11, the first increment of hexagon 6, followed by increments 12, 13, and 14 of hexagon 6. The remainder of increments, 15 through 54 are shown to account for all remaining center increments in hexagons 7 through 13. These increments show that successive radii created from their respective hexagons are intermingled. For example, increment 49 with r squared value 129 (hexagon 13) is smaller than the r squared value of 144 at increment 47 (hexagon 12).

Some hexagons share common radius squared values. Radius squared values of 91, 133, 147, 196, 217, 247, and 259 are framed in bold in Table 1.

Juxtaposed regular hexagons of common edge lengths create a seamless, planar surface, so any vertex of such a juxtaposed grouping could be considered as the center of a "new" pattern, as shown, for example, at the upper left vertex of the fifth hexagon (left) from the original pattern center to the thirteenth hexagon to the right of the original origin.

This results in the eighteenth hexagon at (18 x root 3)/2 (=15.58846 as shown in Figure 12) above the "new" center and an x coordinate r^2 = 243 corresponding to the first radius squared value for hexagon number 18 as well as the 18th hexagon vertex r squared value of 18 x 18 =324 in Table 1. Note from Table 1, the hexagonal increments for even - numbered hexagons are odd numbers, while the hexagonal increments for odd - numbered hexagons are even numbers. The odd number increments for obtaining r squared values along a hexagon segment start with 1 and increase by 2. The even numbered hexagon increments start with 2 and also increase by 2. So, for the "New" origin, its 18th hexagon has an r squared of 18 x 18 =324 and from Table 1, we can determine the r squared value of the next circle center above as 324 – 17 = 307. The one above that is 307 – 15 = 292 and so on as indicated in Figure 12.

Chapter 9

Graphical Derivation of "Missing" Whole Number r Squared Values

Obviously, the flower of life pattern does not explicitly provide the square root of all whole numbers, for example, the square roots of 2, 5, 6, 8, 10, 11, 14, and 15 are not explicitly provided. None the less, the grid pattern can be used along with simple compass – straight edge constructions to plot the associated r squared values as shown in blue in the lower right area of Figure 12. Methods for obtaining "missing r squared values follow:

$r^2 = 2$: Draw a horizontal grid line at Y = -1. The intersection of that grid line with the radius of the circle centered at (0,-1) and connecting a segment from there to (0,0) provides $r^2 = 2$. Then draw the blue circle for $r^2 = 2$.

$r^2 = 5$: Draw a horizontal grid line at Y = -2. The intersection of that grid line with the radius of the circle centered at (0,-2) and connecting a segment from there to (0,0) provides $r^2 = 5$. Then draw the blue circle for $r^2 = 5$.

$r^2 = 6$: Draw a vertical tangent down from where the blue $r^2 = 2$ crosses the x axis to the intersection of the Y = -2 grid line. The segment from this intersection to the origin is $r^2 = 6$. Then draw the blue circle for $r^2 = 6$.

$r^2 = 8$: Draw a vertical tangent down from where the green $r^2= 7$ crosses the x axis to the intersection of the Y = -1 grid line. The segment from this intersection to the origin is $r^2 = 8$. Then draw the blue circle for $r^2 = 8$.

$r^2 = 10$: Draw a horizontal grid line at Y = -3. The intersection of that grid line with the radius of the circle centered at (0,-3) and connecting a segment from there to (0,0) provides $r^2 = 10$. Then draw the blue circle for $r^2 = 10$.

$r^2 = 11$: Draw a vertical tangent down from where the blue $r^2=10$ crosses the x axis to the intersection of the Y = -1 grid line. The segment from this intersection to the origin is $r^2 = 11$. Then draw the blue circle for $r^2 = 11$.

$r^2 = 14$: Draw a vertical tangent down from where the blue $r^2=10$ crosses the x axis to the intersection of the Y = -2 grid line. The segment from this intersection to the origin is $r^2 = 14$. Then draw the blue circle for $r^2= 14$.

$r^2 = 15$: Draw a vertical tangent down from where the blue $r^2= 6$ crosses the x axis to the intersection of the Y = -3 grid line. The segment from this intersection to the origin is $r^2 = 15$. Then draw the blue circle for $r^2 = 15$.

Chapter 10

Expansion by Symmetry

The six around one hexagonal symmetry that occurs in the seed of life about its center circle also occurs around all six of the circles around the original center or to any other circle in the overall pattern. This hexagonal symmetry, as for example, in the right side of Figure 7, can be used to mirror copy this entire pattern about its six outermost hexagon segments as a means to expand the pattern as shown in Figure 13, rather than using the tedious process of "compassing in" individual circles. The characteristic that six equilateral hexagons juxtaposed around the center one occupy a plane with no gaps gives rise to the ability that a pattern center may be designated at the center of any of its unit circles.

Figure 13

Expansion by Symmetrical Mirror Copies

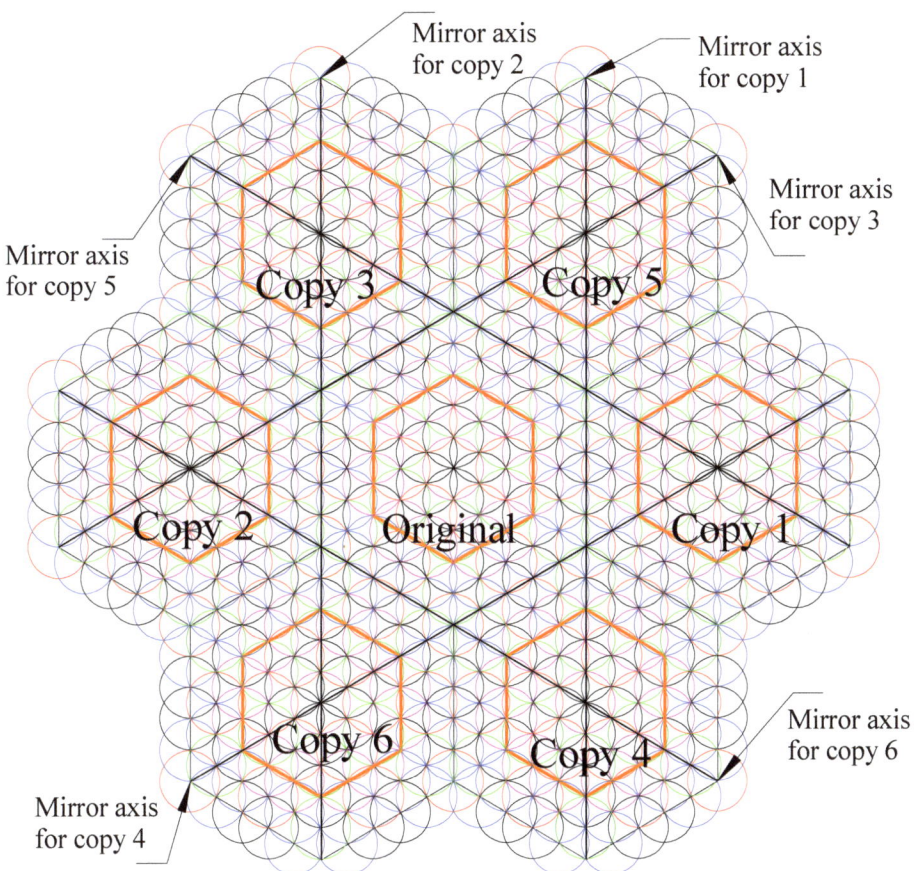

Chapter 11

Summary

The fundamental flower of life formation is started with the vesica piscis and the six-around-one pattern that followed was also created by drawing successive vesica pisces. The unit circle centers of the formation align along the six segments of hexagons starting nearest to the origin with a single frequency (one petal - edged) hexagon, followed concentrically outward with two frequency, then three frequency, four frequency, and so on segment lengths. Each nth hexagon segment has n+1 circles centered on it. The horizontal spacing of petal nodes is ½ root 3 and the vertical spacing of nodes is 1.

The conventionally double – circled "Flower of Life" contains 19 circles, but an additional 36 circles are drawn and then trimmed to create the arcs shown in the left half of Figure 10.

For simplicity, the radius from the origin to each node (petal tip) is squared, indicating the whole number associated with the radius squared and the square root of the numbers at each hexagon vertex (Figures 11 and 12) represents which hexagon is denoted in Table 1.

Any and all distances from any node to another in the flower of life pattern is the square root of a whole number. Using the technique from Chapter 9, it seems intuitively evident that all whole - number square roots could be determined if the pattern were sufficiently extended.

The hexagon increment techniques at the end of Chapter 8 and in Table 1 can be used without physically extending the pattern, which can determine the distance from the origin to all circle centers along the segments of the nth hexagon.

References

1. https://en.wikipedia.org/wiki/Vesica_piscis
2. https://en.wikipedia.org/wiki/Miraculous_catch_of_fish
3. https://www.defendingthebride.com/ss/fish/Fish3Both.pdf
4. https://www.biblegateway.com/passage/?search=John%2021%3A11&version=ESV
5. https://www.outofstress.com/vesica-piscis-meaning/
6. https://i.pinimg.com/originals/d0/f5/63/d0f56349daaff0663f83c84c26ca0a5d.gif
7. https://1.bp.blogspot.com/-gLmKWkbuUGw/Uukj7779zMI/AAAAAAAAvs/-04mn11TerU/s1600/gothic+vp.jpg
8. Crowder, Brent D., The Pyramids of Giza Scaled With Respect to Space and Time, Fulton Books, 2025
   https://www.barnesandnoble.com/w/the-pyramids-of-giza-brent-crowder/1147027225

ABOUT THE AUTHOR

As a teen, Brent enjoyed working for his father, who employed him to install thousands of feet of trenching and piping from waterwells to and within two fifty-acre fields with sprinkler systems. He also assisted in the operation of the sprinkler systems, as well as operation and maintenance of construction and farm and construction equipment such as bulldozers, backhoes, plows, wellhead engines and pumps.

As a student and young man, he experienced more satisfaction studying math, chemistry, thermodynamics, and physics than social studies and history. Math and science gave him the satisfaction of insight into the quantitative cause-effect relations that describe the physics and chemistry of nature.

He earned his bachelor of science degree in mechanical engineering and worked professionally in the power generation industry on design, construction, start-up, and operation on coal-fired, geothermal, natural gas, nuclear, and hydroelectric power plants in the United States.

Through his inherent interest in the mathematics and geometry of nature, he delightfully encountered the "Flower of Life", appreciating the pattern extended from the vesica piscis and "Seed of Life" from which the familiar flower of life image is derived. Then, he discovered that unit (r=1) circles comprising the greater (extended) flower of life pattern have an amazing characteristic, whereby the center – to - center distances between all its circles are always square roots of whole numbers.

Additionally, he concludes the vesica piscis may not be, but perhaps should be more generally appreciated in relation to the reference (4) to the Bible's book of John, the architecture of the Gothic Pointed Arch", and the artistic/religious images that sometimes occur within and around a vesica piscis.

So, Mr. Crowder might be characterized as a "sacred geometry" enthusiast, noting along with others, that a unit radius circle can also serve as a means to graphically obtain the golden ratio (as described in Figure 3 of Ref. 8) and then to a scaled version of the Great Pyramid (Khufu) of Giza, further linking the unit circle and a vesica piscis to a grander geometrically and spiritually connected architecture.

He hopes this book and its connections will inspire others to recognize, appreciate and explore relations of number, geometry, architecture and spirituality.

www.ingramcontent.com/pod-product-compliance
Lightning Source LLC
Chambersburg PA
CBHW040934030426
42337CB00001B/13